Your Step-by-Step Field Guide to Variable Frequency Drives (VFDs)

Install · Wire · Run · Troubleshoot

M. E. SULEIMAN

PREFACE

Over the past three decades, I've worked in a wide range of industrial environments—from high-speed food processing plants to the sophisticated wind tunnels at NASA. I've served as an Electrician, an Industrial Electrical Technician, a Maintenance Manager, and an Electrical Contractor. I've installed, configured, and repaired more VFDs than I can count—sometimes under ideal conditions, but more often in high-pressure situations when systems were down, and the clock was ticking.

One thing has become consistently clear: Variable Frequency Drives are powerful tools—but often misunderstood. Many failures I've seen in the field—costly downtime, damaged motors, erratic system behavior—weren't caused by defective equipment, but by avoidable mistakes in installation, setup, or troubleshooting.

That's why I wrote this book. It's a practical, experience-based guide for those who need to get the job done right.

This guide is brand-agnostic, built around the common principles shared by nearly all VFDs on the market today. Whether you're working with Allen-Bradley, ABB, Schneider, Delta, or any other manufacturer, the knowledge and techniques you'll find here will serve you well.

My goal is to give you a clear, reliable, no-nonsense

resource for installing, operating, troubleshooting, and maintaining VFDs—confidently and correctly.

I hope this guide saves you time, reduces frustration, and helps you build systems that are safer, more efficient, and easier to maintain.

— M. E. Suleiman
S Electric and Controls LLC
California Electrical Contractor License #1120018

INTRODUCTION

In modern industrial environments, Variable Frequency Drives (VFDs) are no longer a luxury—they're essential. From improving motor control and reducing energy consumption to extending equipment life, VFDs are a critical tool for anyone working with electric motors in commercial or industrial applications.

And yet, despite their importance, many professionals still find VFD installation and troubleshooting intimidating. A simple misstep in wiring, grounding, or programming can result in expensive downtime, motor damage, or safety hazards.

This book is your universal guide. It focuses on the foundational concepts, components, and functions that apply across nearly all modern VFDs—regardless of brand. Rather than locking you into one manufacturer's documentation, this guide offers a practical, hands-on understanding of what a VFD is, how it works, how to troubleshoot it, and how to maintain it for reliable performance.

Whether you're a seasoned electrician, a hands-on electrical technician, or a plant maintenance manager overseeing critical systems, this guide is designed with your real-world needs in mind.

Inside, you'll find:

- Step-by-step instructions for safe, efficient VFD installation
- Troubleshooting strategies grounded in real field scenarios
- Wiring diagrams and configuration tips you can rely on
- Insights into fault codes, EMI issues, grounding, and motor compatibility
- Best practices to extend the life of your drives and motors

If you've ever faced a VFD that wouldn't start, a motor behaving erratically, or cryptic error codes, you're not alone. And you're in the right place.

Let's take the guesswork out of VFDs—and get to work.

TABLE OF CONTENTS

CHAPTER 1
WHAT IS A VFD?

A VFD—short for Variable Frequency Drive, also called an AC Drive or Inverter—is an electrical device that controls the speed of an AC motor by varying the frequency and voltage of the power supplied to the motor.

In this guidebook, we'll refer to this device simply as a **VFD**.

Short History

Before the introduction of VFDs, AC motor speeds were changed using mechanical methods like pulley combinations or gearboxes—often with limited speed range and cumbersome adjustments. An alternative was to use DC motors with variable DC drives. However, DC motors had their own issues: carbon brush wear, commutator sparking, limited horsepower capability, and higher maintenance needs.

In the early 1980s, VFDs began appearing in industrial settings, though they were bulky and limited in performance. Thanks to advances in electronics, today's VFDs are compact, affordable, reliable, and widely used across many industries.

How Does a VFD Work?

A VFD receives fixed-frequency AC power (typically 60 Hz), converts it to DC, then inverts it back to variable-frequency AC power. By adjusting the output frequency (0–60 Hz or more), the VFD controls the motor speed accordingly.

For example, if a motor is rated at 1800 RPM at 60 Hz, setting the frequency to 30 Hz reduces the motor speed to 900 RPM.

Where Are VFDs Used?

VFDs are now common in many industries and applications, including:

- Food processing plants
- Conveyor systems
- Packaging lines
- Water and wastewater treatment facilities
- HVAC systems
- Robotics and automation
- Amusement park rides

Anywhere variable-speed motor control is needed, VFDs can improve performance, save energy, and reduce equipment wear.

CHAPTER 2

BASIC (UNIVERSAL) COMPONENTS OF A VFD

All modern VFDs, regardless of brand or size, share a core structure. Understanding this layout will give you the confidence to approach any VFD in the field. Below, we walk through each of the universal components (as illustrated in Fig 1).

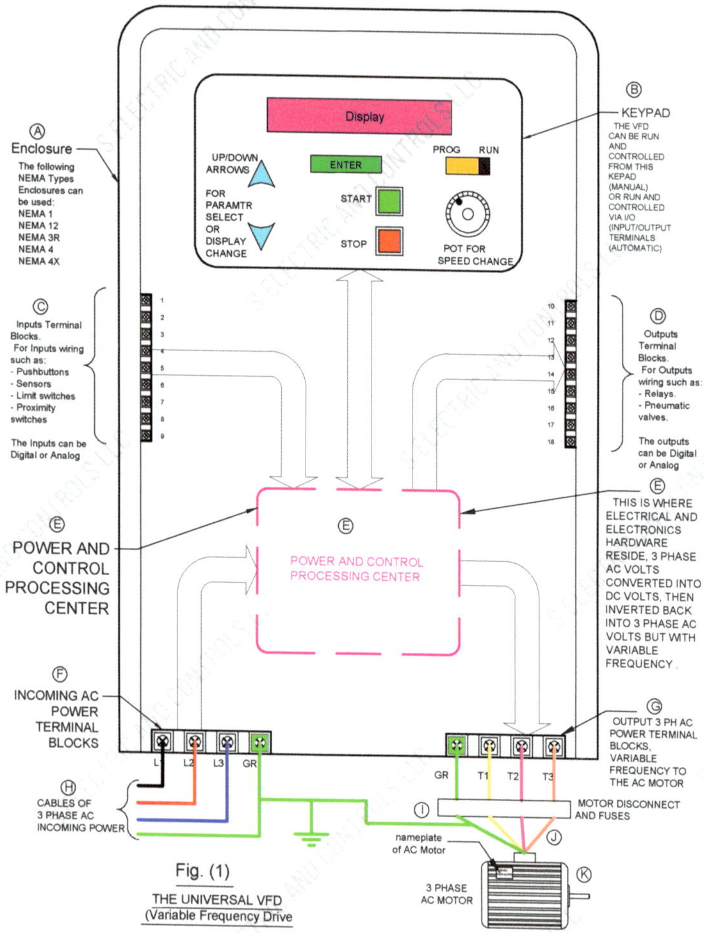

Fig. (1)

THE UNIVERSAL VFD (Variable Frequency Drive

A. Enclosure

VFDs are housed in enclosures rated for their intended environment:

- **NEMA 1**: For clean indoor use (e.g., electrical panels or electrical/mechanical rooms)
- **NEMA 3R**: Sealed against dirt, dust, rain—suitable for indoors/outdoors
- **NEMA 4**: Withstands hose-directed water, splashes, dust
- **NEMA 4X**: Same as NEMA 4, but resistant to corrosive environments
- **NEMA 12**: Dust-tight for indoor industrial use (e.g., workshops, factories)

B. Keypad

The keypad displays frequency, voltage, amperage, and fault codes. You can program the drive, run it, and adjust operating modes through the keypad interface.

C. Input Terminal Blocks

This is where field input signals (digital or analog) are wired in. Inputs can include:

- Pushbuttons
- Sensors (e.g., proximity, limit switches)
- PLC outputs

D. Output Terminal Blocks

These send signals from the VFD to external field devices, including:

- Relays, solenoids (digital)
- Analog control valves
- Signals to PLCs

E. Power and Control Processing Center

This section contains the electronics that:

- Convert incoming AC to DC (via rectifiers)
- Invert DC back to variable-frequency AC
- Monitor and regulate output
- Handle thermal protection and fault generation

F. Incoming AC Power Terminal Blocks

The VFD's main power input terminals. Follow local and national codes during wiring.

G. Output Power Terminals to AC Motor

These terminals connect the VFD to the AC motor. Output frequency and voltage are adjusted according to control inputs.

H. Incoming 3-Phase Power Cable Colors

Standard color codes:

- **Black, Red, Blue** (power phases)
- **Green** (ground)

I. Motor Disconnect and Fuse Block

Fuse block or disconnect switch for protection. Must be sized per NEC and motor rating.

J. Output Cable Colors to Motor

Standard color codes:

- **Brown, Orange, Yellow** (BOY for phases)
- **Green** (ground)

K. AC Motor Nameplate

Crucial for selecting and setting your VFD. Includes:

- Horsepower (HP)
- Voltage
- Full load current (Amps)
- RPM
- Frequency (Hz)
- Frame size
- Number of phases

The motor nameplate is your starting point for correct VFD selection and parameter settings.

CHAPTER 3

HOW TO SELECT A VFD FOR YOUR MOTOR

Selecting the right VFD for your motor involves more than matching voltage and horsepower. Here are the most important considerations:

1. Start with the Motor Nameplate

As covered in Chapter 2(K), the motor nameplate lists:

- HP (Horsepower)
- Volts
- Amps
- RPM
- Frequency (Hz)
- Phase

The VFD must match or exceed the motor's full-load current and horsepower requirements. Undersizing the VFD can result in overheating and poor performance.

2. Understand Your Application

Consider:

- Continuous or intermittent duty
- Constant or variable load
- Start conditions (e.g., high initial torque)
- Load characteristics and operating cycles

3. Type of Load

What kind of equipment will the motor drive?

- Centrifugal pumps (high-volume, constant-torque)
- Conveyor belts (variable loads)

- Shock or impact loads

Your VFD should be rated for at least 110% of the motor's full-load current.

4. Environment

Where will the VFD be installed?

- Clean or dusty?
- Dry or humid?
- Hot ambient temperatures?

Select the appropriate NEMA enclosure and cooling strategy.

5. Evaluate Reliability, Performance, and Price

Balance your selection between cost, durability, ease of programming, and available features.

6. Installation and Maintenance Matters Too

Even the right VFD will underperform if not installed correctly. Proper parameter setting and ongoing preventive maintenance are crucial. These are discussed in detail in the chapters that follow.

CHAPTER 4
INSTALLATION AND WIRING OF VFD

Installing Your New VFD

1. **Heat Considerations**: VFDs generate heat and usually have built-in fans, but they still require adequate spacing if installed in control panels. Leave sufficient ventilation space.

2. **Installation Location**: Choose a dry, clean indoor environment, free of corrosive vapors. Maintain a minimum of 42 inches of clearance in front of the panel for service.

3. **Protection from Debris**: If the VFD is mounted in the lower part of a cabinet, install a protective shield (with ventilation space) to prevent falling debris from entering the unit.

4. **Mounting Practice**: Mount the VFD square and flush with the backplate—no washers or spacers. The backplate also acts as a heat sink, so ensure good contact.

5. **Accessibility**: Install the VFD where it can be easily reached for inspection and service.

6. **Environmental Protection**: Keep the VFD away from moisture and direct sunlight.
7. **Follow OEM Instructions**: Always consult the specific installation guidelines provided by the VFD manufacturer.

Wiring Your VFD

Before wiring:

- Have your schematic drawings ready.
- Size and label all cables.
- Use correct conductor types as per NEC and local codes.

You will typically deal with two categories of wiring:

1. **Power Wiring** (line and motor connections)
2. **Control Wiring** (start/stop signals, analog inputs, etc.)

Keep these separate to minimize electromagnetic interference (EMI).

EMI Tip: Shield control cables and ground the foil shield on one side only to avoid ground loops.

A. Input 3-Phase AC Power Wiring

- Terminals: L1, L2, L3 (plus ground)
- **Sizing**: Per NEC 2023 430.122(A), conductors

must be rated for at least 125% of the VFD's input current (as listed on its nameplate).

- Use NEC Table 310.16 to select the proper conductor size.

B. Output Power Wiring to AC Motor

- Terminals: T1, T2, T3 (plus ground)
- **Sizing**: Per NEC 2023 430.122(B), conductors must be rated for at least 125% of the motor's full-load current (FLC).
- Again, use NEC Table 310.16, factoring in insulation, ambient temp, and cable bundling.

Grounding and Bonding

- Ground and bond both VFD and motor properly.
- **Grounding** provides a low-impedance path for fault current.
- **Bonding** equalizes potential between conductive parts.
- Consult NEC Articles 250 and 430.130 and your OEM's manual.

C. Control I/O Wiring

- Inputs: Digital and analog signals to control the VFD.

- Outputs: Status or control signals sent from the VFD to other devices.

Refer to **Figure 2** and **Figure 3** for sample wiring:

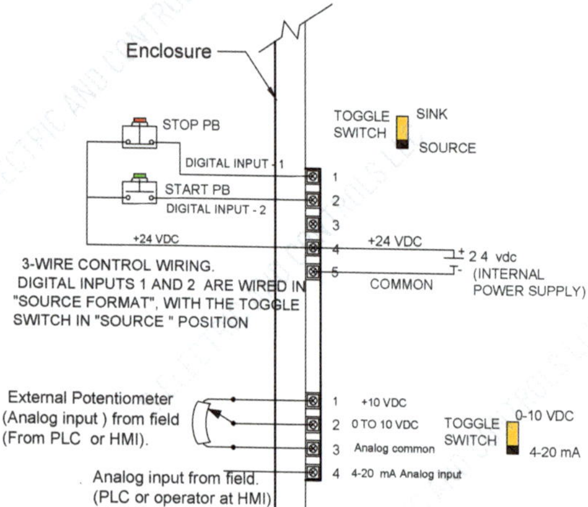

Fig. (2)

Input wiring - SOURCE DIGITAL AND ANALOG - EXAMPLES

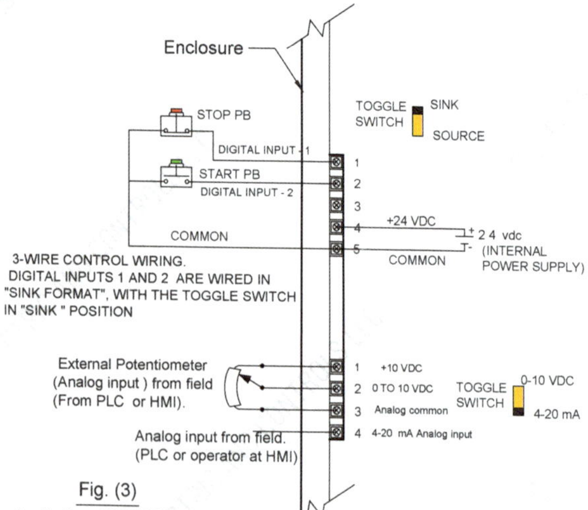

Fig. (3)

Input wiring - SINK DIGITAL AND ANALOG - EXAMPLES

Figure 2 (SOURCE Mode):

- Digital Inputs #1 and #2 use +24V as common (wired to terminal #4).
- Toggle switch is set to SOURCE mode.

Figure 3 (SINK Mode):

- Digital Inputs #1 and #2 use -24V common (wired to terminal #5).
- Toggle switch is in SINK mode.

Analog Inputs:

- Terminals #6, 7, 8 for Potentiometer (POT)
- Terminal #9 for 4–20 mA signal from PLC or other device
- Select the source (POT or analog) in the VFD parameters.

D. Final Wiring Checklist

1. Update all schematic drawings with field changes.
2. Verify tightness and correct placement of all connections.
3. Clearly label and organize all wires.
4. Separate power and control cables to minimize EMI.

5. Follow all safety procedures during installation and testing.

Once installed and wired, you're ready to move on to parameter programming and commissioning.

CHAPTER 5
PARAMETERS – SETTINGS AND PROGRAMMING

After installation and wiring, it's time to program the VFD parameters to match your motor and application needs.

Safety Note: You may apply AC power to the VFD for programming while the motor cables are disconnected at the VFD terminals. Use protective gear and safe practices.

What Are VFD Parameters?

Parameters are settings that define how the VFD will run the motor—its behavior, speed, input source, limits, and protection settings. Each manufacturer provides a list of available parameters and their functions.

Programming Methods

You can set parameters:

- Through the front keypad (most common)
- Via software using a laptop
- Via communication protocols (Modbus, Ethernet/IP, etc.)

Tip: The keypad method is practical and easy to learn. Follow on-screen prompts.

Common Parameters (Example values shown)

Param #	Description	Example Value
1	Motor Nameplate Voltage	460 VAC
2	Motor Frequency	60 Hz
3	Motor Full Load Current	12 Amps
4	Minimum Frequency	0 Hz
5	Maximum Frequency	60 Hz
6	Start Command Source	Keypad / 2-wire
7	Stop Mode	Coast / Ramp
8	Speed Reference Source	POT / 4–20mA
9	Acceleration Time	10 seconds
10	Deceleration Time	10 seconds

Refer to your motor nameplate and match these settings carefully.

Advanced Parameters

1. **DC Braking**: Useful for precise stops (e.g., conveyors or vertical loads)
2. **Torque Boosting**: Helps with torque at low speeds
3. **PID Control**: For applications requiring feedback-based control (e.g., pumps, temperature regulation, mixing ingredients proportionally per specific recipe' in Food Processing Plants).

   *(See **chapter 9** for PID simply explained).*

Advanced parameters fine-tune performance. Use them only if the basic configuration isn't sufficient.

Consult the OEM manual for access and descriptions of advanced features.

Next step after programming is to begin testing and commissioning, covered in Chapter 6.

CHAPTER 6

TESTING, RUNNING AND COMMISSIONING YOUR VFD (FIRST TIME)

Once you've completed the installation, wiring, and parameter programming of your VFD, follow these steps to test and commission the system safely and effectively:

1. Do Not Apply Power Yet

- Keep the system locked out and tagged out.
- Conduct a pre-check: Ensure updated schematic drawings, VFD manuals, PPE (Personal Protective Equipment), and relevant documents are ready and placed in the panel's document pocket.
- Post the OEM fault code sheet visibly (but not blocking any cooling vents).

2. Conduct a Thorough Visual Inspection

- Verify tightness of all electrical terminations.
- Inspect fasteners, wire harnesses, grounding points.

- Ensure cable labeling is clear and numbers face outward.

3. Grounding Continuity Check

- Use a multimeter to confirm good continuity across all grounding and bonding points.

4. Apply Power for Dry Test (Without Load)

- If possible, decouple the motor shaft from the load or disconnect motor leads at the VFD.
- Apply power carefully with PPE.
- Check the VFD display. Once powered, review all programmed parameters against your documented settings.

5. Reconnect Motor Leads

- Once the parameter check is complete, disconnect power again.
- Reconnect the motor cables to the VFD output terminals.

6. Verify Inputs and Communications

- Check the functionality of sensors, control devices, and any communication links.
- Confirm that all permissive signals are satisfied (i.e., ready-to-run conditions are met).

7. Verify Motor Rotation Direction

- Set a low-speed command using the keypad or wired controls.
- Jog the motor briefly and observe the rotation.
    - For bidirectional applications, define the default direction.
    - For unidirectional needs, reverse direction (if necessary) by:
        * Swapping any two motor wires at the output terminals
        * Or adjusting wiring/program settings (e.g., 3-wire Forward/Reverse + parameters).

8. Perform Dry Run

- Start the system at low frequency and gradually increase speed.
- Observe mechanical movement and system response.
- Run for 30+ minutes, monitor:
    - VFD display values (voltage, current, frequency)
    - Motor behavior
    - System noise and vibration
- Log voltage/amperage readings phase-to-phase.

9. Post Test Review

- Monitor for fault codes or warnings.
- Check temperatures of VFD and motor.
- Adjust parameters if needed for smoother performance.

10. Final Commissioning

- Handover system to the production team, application operator, or end user.
- Test the drive under actual load conditions.
- Push to operating limits and observe:
  - Acceleration/Deceleration smoothness
  - Overload behavior
  - Nuisance tripping or parameter thresholds

Final Tip:

Real-world testing is key to tuning parameters effectively for reliable long-term performance.

Your system is now commissioned. Chapter 7 covers troubleshooting techniques for resolving issues that may arise during operation.

CHAPTER 7
TROUBLESHOOTING

Troubleshooting is the process of collecting facts and using logical thinking to isolate and resolve equipment issues. With VFDs, methodical troubleshooting saves time, reduces downtime, and prevents unnecessary equipment damage.

Basic Steps in Troubleshooting Your VFD

1. Know Your Equipment

- Spend time familiarizing yourself with the VFD and the connected system.
- Study the OEM manual, wiring diagrams, and control logic.
- Understand the purpose and operation of each component.

2. Start with the Obvious

- Check power sources, breakers, and fuses before digging into complex issues.
- Visually inspect connections, cable damage, and environmental conditions.

- Check the VFD Display.

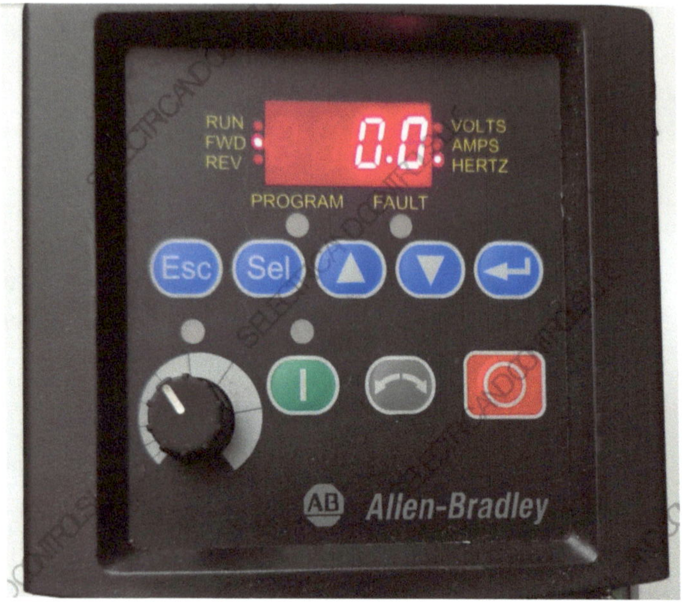

3. Talk to the Operator

- Operators often notice subtle changes in performance.
- Gather system history, symptom timelines, and any recent changes in load or controls.

4. Consult the OEM Fault Codes List

- Most VFDs display fault codes during failure events.
- Refer to the fault list in the OEM manual or label on the side of the VFD.

- Example: Figure 4 (Allen-Bradley PowerFlex) includes a basic fault code list and parameter summary.

5. Use a Handheld Signal Generator (Fig. 6)

- For analog input faults (e.g., loss of 4–20 mA signal from a PLC), simulate the signal using a signal generator.
- This tool helps verify if the VFD is receiving and responding to analog inputs properly.

6. Overtemperature Faults

- Possible causes: blocked airflow, dirty filters, failing cooling fan.
- Clean out the VFD enclosure, replace damaged fans, or restore ventilation.

7. Overload Faults (see Fig 7 as example of list of Faults Codes)

- Causes: excessive mechanical load, tight bearings, improper parameter settings, or a failing motor.
- Check motor amperage, shaft movement, and verify correct acceleration times.

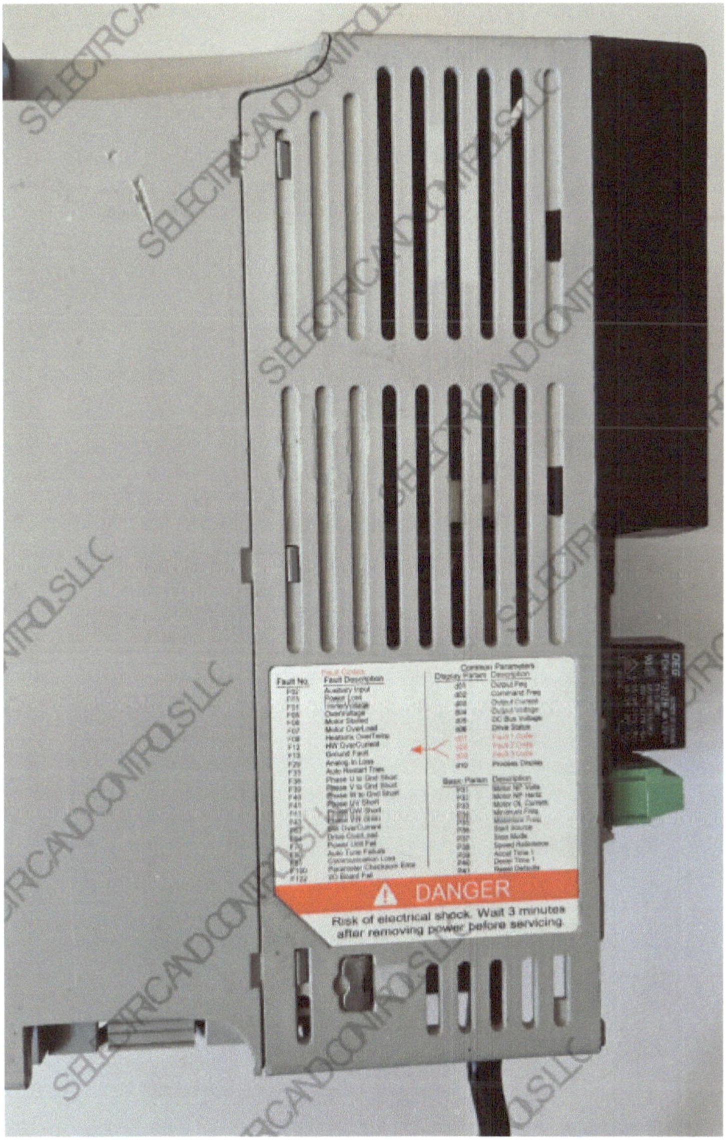
DANGER
Risk of electrical shock. Wait 3 minutes
after removing power before servicing.

8. Startup Belt Screeching

- If the VFD is used with V-belt-driven pulleys (e.g., chillers or conveyors), a screeching noise may indicate belt slippage.
- First check mechanical issues like belt tension, then consider increasing acceleration time in the VFD parameters.

9. Overvoltage Faults

- May occur due to sudden load drops, regenerative braking, or long deceleration times.
- If it resets and runs fine, it might have been a transient condition.
- If recurring, consider enabling dynamic braking or adjusting deceleration parameters.
- Repeated overvoltage could indicate aging or overstressed internal components.

10. Don't Open the VFD Internals

- OEMs do not recommend opening or repairing the internal electronics of a VFD in the field.
- Contact the manufacturer's support team for internal failures.

Tip: Keep a logbook of all troubleshooting steps and findings. Over time, this becomes a valuable reference for maintaining long-term system reliability.

Chapter 8 will focus on preventive maintenance practices that can reduce the need for frequent troubleshooting and extend the life of your VFD.

CHAPTER 8

PREVENTIVE MAINTENANCE FOR YOUR VFD

Like any critical piece of industrial equipment, a Variable Frequency Drive (VFD) requires regular preventive maintenance to ensure long-term reliability, reduce the risk of unplanned failures, and maintain peak performance.

Key Preventive Maintenance Tasks

1. Maintain a Clean Operating Environment

- Keep the VFD enclosure and surroundings free of dust, debris, and foreign objects.
- Dust buildup inside the VFD or on its filters reduces heat dissipation and can cause overheating.

2. Inspect and Maintain Cooling Fans

- Listen for unusual fan noises or vibrations—signs of worn bearings or damaged blades.
- Replace or repair faulty fans immediately.

3. Control Moisture and Humidity

- Ensure the VFD is installed in a dry location.
- Moisture combined with dust can form conductive sludge or cake layers inside the VFD, affecting cooling and electronics.

4. Retorque Cable and Terminal Connections

- Over time, thermal expansion and vibration can loosen cable terminations.
- Include torque checks as part of your routine maintenance schedule.

5. Monitor Internal and Ambient Temperatures

- Use temperature sensors or VFD diagnostics to monitor internal cabinet temperatures.
- Take corrective action if operating temperatures approach VFD limits.

6. Use a Structured Maintenance Checklist

- Fig. 7 shows a sample monthly maintenance checklist.

S Electric and Controls, LLC
123 street X, Any City, California, USA

MAINTENANCE DEPARTMENT

PREVENTIVE MAINTENANCE SCHEDULE

Electrical Technician name: John Doe **Department: Maintenance -** Electrical **Shift:** 1st

Date: 6/16/2025 **Equipment Name:** VFD for the Chiller **Serial #:** 1234JK567 **Freq:** Monthly

Task #	Task Description	Parts used	Done
1	Safety: Lockout/Tagout VFD and Chiller Power System.		
2	Clean the VFD from Dust and Debris: clean/vacuum the VFD control cabinet.		
3	Take out the front cover of VFD. Clean/vac inside the VFD, cooling fans, and through the enclosure.		
4	Clean/vac or replace the air filters. Put back the front VFD cover.		
5	Visually inspect the Control cabinet housing the VFD for any signs of moisture. Check the Door's seals.		
6	Inspect and retighten all cables screws and fasteners. Check thoroughly all wirings/groundings.		
7	Reinstall all removed components, control covers, wire ducts' covers and make final visual inspection.		
8	Remove the lockout/Tagouts, use your PPE and test-run the VFD.		
9	Take readings of incoming volts/currents, output volts/currents to the AC Motor, monitor all.		
10	Log all readings, shutdown properly the VFD and the system before turning it over to the Production Dept.		
11	Complete and sign this form with the date of completion.		

NOTES:

Electrical Technician name ----------------------------- Signature -----------------------------
Date of completion of this Preventive Maintenance Schedule / /

Name of Supervisor: ---------------------------------- Signature: -----------------------------
Date : / /

Comments:

- Create a customized preventive maintenance program for your plant that includes:
    - **Daily** checks (visual inspections, error logs)
    - **Monthly** tasks (filter cleaning, connection checks)
    - **Semi-Annual/Annual** inspections (parameter reviews, fan replacement, full internal inspections).

Tip: Keep a maintenance logbook or software system to record inspections, findings, and completed tasks. Patterns in maintenance data often help predict future failures before they happen.

With preventive maintenance in place, your VFD is more likely to deliver long-term, trouble-free operation.

CHAPTER 9

PID CONTROL – SIMPLY EXPLAINED

A Variable Frequency Drive (VFD) can run an AC motor smoothly, efficiently, and with significant energy savings. But in automation systems where **precision** and **fast response** matter, **PID control** takes performance to the next level.

What is PID?

In process control, you often want a system to reach and maintain a desired condition — such as **temperature**, **speed**, or **flow** — despite changes in load or disturbances. This is where PID control comes in.

Key terms:

- **SP (Set Point)** – The target value for the process (e.g., 60 Hz motor speed, 75 °F temperature, or 100 L/min flow rate).
- **PV (Process Variable)** – The current measured value of the process (feedback from sensors).
- **PID (Proportional–Integral–Derivative)** – A control algorithm that adjusts output based on three factors: **P**, **I**, and **D**.

The Three PID Gains:

1. P – Proportional Gain (Kp)

- Responds to the current error between SP and PV.
- A higher Kp means a stronger correction, but too high can cause overshoot and oscillation ("hunting").

2. I – Integral Gain (Ki)

- Responds to the *accumulated* error over time.
- Helps eliminate steady-state error, but too high can cause slow oscillations around the set point.

3. D – Derivative Gain (Kd)

- Responds to the *rate of change* of PV.
- Predicts where the process is heading and applies correction before the error grows.
- Helps smooth the response and reduce overshoot.

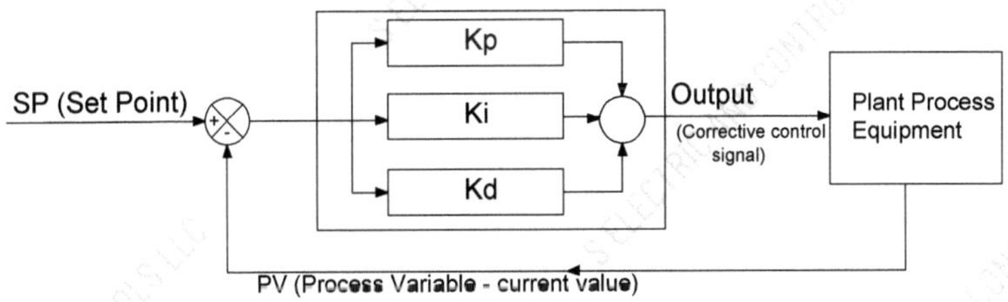

FIG. 9

PID Controller System

How PID Works

1. The **operator** enters the **SP** and sets gain values (Kp, Ki, Kd) manually or uses auto-tuning.
2. The **controller** continuously compares **PV** to **SP**.
3. Based on P, I, and D calculations, the controller sends corrective output signals to the VFD.
4. The VFD adjusts motor speed or torque to keep PV close to SP.

Tip: Some applications work best with only **P + I**. Others need all three terms. PID tuning is part science, part art — and it varies by process.

Tuning PID Gains

- **Manual Tuning** – Adjust Kp, Ki, and Kd one at a time while observing the system's response.
- **Auto-Tuning** – Many modern VFDs can automatically calculate and set optimal gains for faster commissioning.
- **Rule of Thumb** – Increase Kp until the system responds quickly but without instability, add Ki to remove steady-state error, and use Kd to smooth the response.

Real-World Example: Food Processing Plant

Application: Coating cereal flakes with syrup in precise proportion.

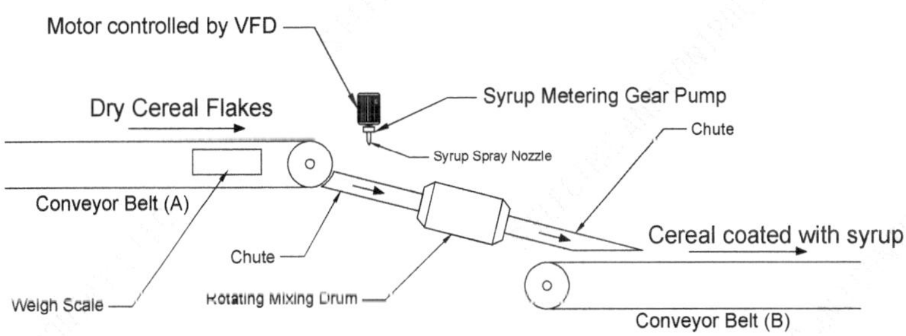

Fig. 10

Coated Cereal Flakes with Syrup
(Syrup Pumping controlled by VFD with PID Parameters)

Process flow (Fig. 10):

1. **Conveyor Belt A** moves dry cereal flakes over a **weigh scale**.
2. The scale measures the weight (PV) and sends the data to the VFD.
3. The VFD controls a **syrup pump**, releasing syrup through a spray nozzle in exact proportion to the cereal weight (SP: 1 oz syrup per 1 lb flakes).
4. Coated flakes move into a **mixing drum**, then onto **Conveyor Belt B**.

Why PID is critical here:

- If the cereal flow fluctuates, the syrup output must adjust instantly to maintain the recipe ratio.
- **Without PID:** The syrup flow may lag behind changes, causing dry spots or overcoating.
- **With PID:** The VFD reacts quickly and smoothly, keeping coating consistent.

CHAPTER 10

CONCLUSION: MASTERING VFDS THROUGH KNOWLEDGE AND PRACTICE

This guide has taken you on a step-by-step journey through the world of Variable Frequency Drives (VFDs) — from understanding what they are and how they work, to installing, programming, troubleshooting, and maintaining them in real-world industrial environments.

Over the course of these chapters, we covered:

1. What a VFD Is and Why It Matters

- VFDs control the speed and performance of AC motors, reducing energy costs, improving process control, and extending equipment life.
- A solid grasp of their history and evolution helps appreciate today's advanced, compact, and highly reliable designs.

2. Main Components of a VFD

- Knowing the universal layout (enclosure, keypad, inputs/outputs, power section) removes the fear factor when approaching any brand or model.

- Understanding terminal layouts and signal types is key to correct wiring and integration.

3. How to Select the Right VFD

- Motor nameplate data is the starting point.
- Load type, environment, duty cycle, and application demands guide correct sizing and selection.

4. Installation and Wiring

- Follow NEC codes, manufacturer guidelines, and best practices for cable sizing, grounding, EMI reduction, and clean, well-organized wiring.
- Physical installation should allow proper cooling, easy access, and protection from environmental hazards.

5. Parameter Settings and Programming

- Parameters translate application requirements into VFD behavior.
- Understanding key settings such as start/stop sources, acceleration/deceleration, and speed references ensures optimal performance.

6. Testing, Running, and Commissioning

- Safety-first approach: lockout/tagout, verify connections, check grounding.

- Dry-run testing helps confirm parameter settings before load testing under real conditions.

7. Troubleshooting

- Combine observation, operator input, OEM fault codes, and logical diagnostics to pinpoint issues.
- Use tools such as signal generators for analog input testing, and always check simple causes before diving deeper.

8. Preventive Maintenance

- Cleanliness, cooling fan health, moisture control, and retorquing connections are essential.
- Regular maintenance schedules prevent downtime and extend VFD life.

9. PID Control for Precision Applications

- Proportional, Integral, and Derivative gains fine-tune system response for stable, accurate control in demanding processes.
- Practical examples, like controlling syrup coating on cereal flakes, show how PID can prevent product variation.

Final Takeaway

Working with VFDs is a balance of **technical knowledge, safety discipline, and hands-on skill**.

- **For beginners**, this guide provides a clear path to confidence in installation and operation.
- **For experienced technicians**, it serves as a reference and refresher, reinforcing best practices and offering field-proven tips.

By applying what you've learned here — and by staying curious, methodical, and safety-focused — you'll be equipped to keep VFD-driven systems running reliably, efficiently, and profitably for years to come.

Remember: A VFD is not just a black box — it's a vital part of your plant's heartbeat. Treat it with respect, keep it well-maintained, and it will serve you well.

VFD QUICK REFERENCE GUIDE

(For Electricians, Industrial Technicians, and Maintenance Managers)

1. Before You Begin

- **Lockout/Tagout** and follow all safety protocols.
- Review OEM manual and schematic drawings.
- Gather PPE, tools, and test equipment.

2. Selection

- Match VFD HP/kW to motor nameplate or next higher rating.
- Verify voltage, current, and application type.
- Consider environment, cooling, and load type.

3. Installation & Wiring

- Keep spacing for cooling; avoid moisture/dust.
- Size cables per NEC 430.122 and Table 310.16.
- Separate power and control wiring; shield control cables (ground one end only).
- Proper grounding & bonding — NEC 250, 430.130.

4. Parameter Programming

- **Common Parameters**:
  - Motor volts, frequency, full-load amps.
  - Min/Max frequency.
  - Start/stop source & speed reference.
  - Accel/decel times.
- Adjust per application; verify against motor nameplate.

5. Testing & Commissioning

- Dry-run test with motor uncoupled or disconnected.
- Verify motor direction at low speed.
- Log voltage/current readings.
- Adjust parameters for stable performance.

6. Troubleshooting

- Check obvious issues first (breakers, fuses, wiring).
- Use OEM fault codes & operator feedback.
- For analog issues, simulate input with signal generator.
- Overheating? Inspect fans, vents, and load.

7. Preventive Maintenance

- Keep VFD and environment clean/dry.
- Inspect & replace cooling fans as needed.
- Retorque electrical connections periodically.
- Monitor temperature & maintain logs.

8. PID Control Basics

- **P**: Proportional — immediate correction.
- **I**: Integral — eliminates steady-state error.
- **D**: Derivative — predicts and smooths changes.
- Tune Kp, Ki, Kd for stability and responsiveness.
- Use auto-tuning if available.

Pro Tip: Always document changes, maintain updated schematics, and train operators on the basics of VFD use and monitoring.

www.ingramcontent.com/pod-product-compliance
Lightning Source LLC
LaVergne TN
LVHW051136210926
848482LV00006B/333
* 9 7 9 8 3 4 9 6 5 8 1 1 2 *